SOCIAL HOUSING

HOMES IN 25 PAINTINGS

THURSTON JONES

Contents

Coming Home to an Almshouse 4

Coming Home to a Philanthropic Home 6

Coming Home to a Back-to-Back Home 8

Coming Home to a Garden City Home 10

Coming Home to a Home Fit For Heroes 12

Coming Home to a Solid Wall Home 13

Coming Home to a Non-Traditional Home 14

Coming Home to a Temporary Bungalow 16

Coming Home to a Metal Framed Home 18

Coming Home to a Precast Concrete Home 20

Coming Home to an In situ Concrete Home 22

Coming Home to a Timber Framed Home 24

Coming Home to a Concrete Block Home 26

Coming Home to a Cavity Wall Home 28

Coming Home to a Maisonette 29

Coming Home to a Concrete High-Rise Home 30

Coming Home to a Brick Clad, High-Rise Home 32

Coming Home to a Low-Rise Home 34

Coming Home to a Rationalised Traditional Home 36

Coming Home to a Flat Roofed Home 38

Coming Home to a House Alteration 40

Coming Home to a House Conversion 42

Coming Home to a Modern Extra Care Home 44

Coming Home to a Modern Methods of Construction Home 46

Coming Home to a Modern Timber Frame Home 48

'Coming Home to an Almshouse.'
(990 to present day)

Originally called hospitals, these homes date back to the 10th century and were places of residence for the elderly. Wealthy benefactors from society with a social or spiritual conscience came forward to found many of these establishments throughout the UK. Typically, the benefactors were members of the monarchy, the church, the aristocracy or wealthy merchants.

These homes offered subsidised rental for those who could not afford standard rates.

Above: Early solid stone almshouse.

Right: Early timber frame almshouse.

'Coming Home to a Philanthropic Home.'
(19th Century)

Victorian and Edwardian philanthropic housing was mainly provided by wealthy industrialists who wanted better living conditions for their workers.

Some philanthropists demolished slum areas and built new homes for their workers. Others bought existing buildings in slum areas and modernised them by providing greater space, better sanitation and communal facilities. Communal areas boasted schools, halls and allotments.

Philanthropic ideas continued with moving the factory to a rural area to provide healthier living conditions for their workforce.

These philanthropists started the first social housing movement which spread throughout Britain and was duplicated by many other countries.

From the late 19th century, local authorities became the main builders of social housing properties.

Right: Early solid wall homes with communal areas.

'Coming Home to a Back-to-Back Home.'
(Late 18th Century)

Back-to-Backs are predominantly a solid wall form of terraced housing which shares a party wall on three of their four sides.

Thousands of these homes were built during the Industrial Revolution to accommodate the vast populations of thriving factory towns and cities.

Consequently, these terraces became associated with overcrowding, poor health and slums. The houses could be three storeys high and house up to fifteen families in each house.

The Government embarked on the first comprehensive plan to build social housing and demolish these densely populated, privately rented properties.

Right: Early solid wall buildings

'Coming Home to a Garden City Home.' (1898 onwards)

The Garden City Movement tried to incorporate the benefits of the countryside environment with the town environment whilst trying to design out the disadvantages in both within a society.

It was the vision of Ebenezer Howard and was an alternative to the squalor caused by the Industrial Revolution. Through better housing, better comunities and better union between town and country, Howard suggested that a better civilisation could be created.

This movement was to be adopted by councils as they were encouraged to develop new estates and to follow the garden city principles.

The Garden City Movement has been considered revolutionary ever since, as the concepts are based on principles such as collective land ownership, affordable rents and the importance of residents' wellbeing mixed with nature. It became one of the bedrocks for modern sustainable development.

Right: Solid wall with render finish

'Coming Home to a Home Fit For Heroes.'
(1919 onwards)

After World War I, returning soldiers needed good quality homes and affordable housing. Lloyd George promised to provide 'homes fit for heroes.'

The Housing and Town Planning Act (Addison Act) 1919 was passed to create a temporary measure to meet the housing need. It made local councils responsible for assessing local housing needs and planning how to meet them. Local councils started to build the first council housing.

'Coming Home to Solid Wall Home.'
(Common 1800 and the 1940s)

Solid wall homes are constructed of one skin of masonry which is typically brick or blockwork and does not have a designated cavity between the exterior and interior.

The Housing Act of 1930 encouraged slum clearances and local councils started to build new homes using this traditional build method.

Above:RHS semi-detached has been later modernised with external wall insulation.

'Coming Home to a Non-Traditional Built Home.' (World War I Onwards)

During World War II, factories and labour throughout Britain were adapted to assist the war effort. Also, due to heavy bombing, nearly half a million homes were destroyed or structurally damaged. Britain needed to build homes fast and efficiently.

After the war, these factories and their labour had to diversify to survive. To support the post-war recovery, non-traditional methods of construction and prefabrication home systems were embraced.

There are hundreds of types of non-traditional construction built in Britain between 1919 and 1976.

Right: a non-traditional home.
Mansard roof- timber built.
Ground floor- concrete components.

'Coming Home to a Temporary Bungalow.'

Following the Blitz, the Temporary Housing Programme 1943 was passed and offered instant temporary housing. The mass production of these 'prefabs' or part fabricated houses was done in factories to a high standard and then delivered to the site.

Many of these single-storey temporary bungalows had their internal and external components assembled in factory conditions.

There were four main types of temporary bungalows manufactured after the war including steel framed, timber framed and the aluminium alloy which was manufactured from recycled aircraft.

They were welcomed by the homeless and some are in existence today with a Grade II status.

Right: LHS bunalow has been later modernised with brick wrap.

'Coming Home to a Metal Framed Home.'
(Non-Traditional)

There was a surplus of steel production facilities due to the tank and weapon production for the war. One of the non-traditional methods was metal framed construction.

These builds came in a variety of shapes and sizes and were made, and partly assembled within a factory environment. This allowed for a high quality in control during the production of each component.

These homes were often modernised in the early nineties which involved various forms of wall insulation including drilling through the metal sheet attached to the first floor and adding external insulation to the ground floor.

Right: Metal framed home later modernised with insulated render on ground floor.

'Coming Home to a Precast Concrete Home.'
(Non-Traditional)

Non-traditional methods such as precast reinforced concrete (PRC) construction came to the fore at a time of skilled labour and material shortage in post-war Britain.

The structural concrete columns and cladding panels were made off-site before being fixed into place on site. These concrete components were reinforced with steel rod; bolted together or constructed onto a steel frame.

Twenty-two types of PRC structures were deemed defective under the Housing Defects Act 1984 due to corrosion of the metal reinforcing steel in their concrete structure.

Thousands of these homes are still in existence and the majority of them have been externally insulated or brick wrapped.

Right: Semi-detached PRC homes

'Coming Home to an In situ Concrete Home.'
(Non-Traditional)

Concrete was considered a viable building material for domestic housing at the time. The formwork was constructed on-site or in situ, with steel reinforcements added. Liquid concrete was poured into the formwork to construct the structures and sub-structures.

The coarse concrete did not contain a high quantity of a fine aggregate and was referred to as 'no-fines'. This created a higher insulation value and became more thermally efficient when used with cavity wall production.

Right:LHS semi-detached has been modernised with external wall insulation.

'Coming Home to a Timber Framed Home.'
(Non-Traditional)

These homes were imported from various European countries in kit form, to be built on-site in Britain. They were imported in prefabricated sections to offer affordable homes and are considered to be a non-traditional build.

Typically, these builds comprised of single-storey timber frame panels that were boarded internally. The external skin could be clad with vertical tile, brickwork or render.

Many of these homes were constructed entirely of timber. They consisted of masonry chimney stacks to supply the heating and hot water throughout.

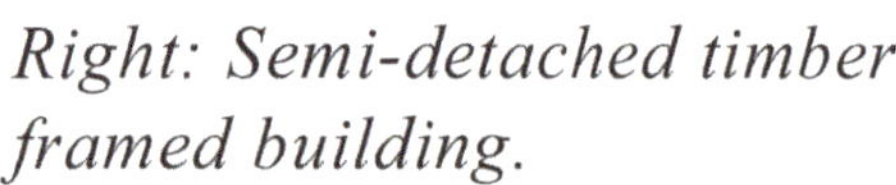

Right: Semi-detached timber framed building.

'Coming Home to a Concrete Block Home.'

These traditional built homes substituted the standard 9" house brick with a more accessible material - concrete. These concrete blocks were made in a larger size to allow for quicker construction.

This was considered a new method of construction as it also reduced the skills required to lay these larger blocks.

Cavity wall homes started to replace solid brick or block homes and cavity wall coincided with a massive post-war, council house building program.

Mansard Roofs

Mansard Roofs are distinctive in that they are double pitched and they have steep sides. The shape of the roof allows for more internal space.

Right:Block house with mansard roof

'Coming Home to a Cavity Wall Home.'

Cavity wall construction was typically built from 1935 up to the modern day and is considered a traditional build method. Research revealed that a cavity between the inner and outer walls of a building would retain the heat and repel dampness better than a solid wall.

Many of these properties were later cavity filled with insulation but early designs had narrower cavities and were difficult to insulate due to excessive solid mortar in them.

'Coming Home to a Maisonette.'

A maisonette is a two-storey flat with direct walk-in access. Maisonette or 'small house' offers housing over garages; shops and other maisonettes and are predominately cavity build structures. They were introduced in the 1960s and 1970s as another way of contributing to higher density housing in the suburbs. The properties do not have an internal communal area.

'Coming Home to a Concrete High-Rise Home.'

Throughout the 1950s and 1960s, prefabricated high-rise blocks started being constructed. The Government encouraged developers to build structures higher than six storeys by providing subsidies as an incentive. These replaced the slum areas of poor quality back-to-back houses.

The high-rise blocks were designed with external walkways and balconies and were referred to as 'streets in the sky'. The maintenance and refurbishment of these blocks became challenging due to their design. It was often more financially viable to demolish the blocks and build new ones.

Left: Modernised high-rise blocks with structural, external wall insulation.

'Coming Home to a Brick Clad, High-Rise Home.'

Traditional housing construction influenced these builds. The vast majority of residential high-rise buildings are built of reinforced concrete. Some very early buildings used brick and concrete combinations. Concrete floors were built on supported columns. Steel reinforcing bars were embedded within the concrete slabs to make the components stronger.

Brickwork was used as the outer wall or cladding and blockwork was used for the inner wall. The load is only bearing of its own weight, not the entire building.

Typically, these were modernised with cavity insulation but many were externally insulated with structural cladding.

At the time, the main reason for the demand for tall buildings was the larger amount of space for homes within the floor footprint for each block. This allowed for a greater population density of residents and as a result, created cost savings for developers.

Right: As-built brick clad, high-rise.

'Coming Home to a Low-Rise Home.'

As neighbourhoods all over Britain were being demolished and rebuilt, the town planning encouraged the building of both high-rise and low-rise blocks of flats.

Typically, these were traditional cavity built structures with an internal communal area accessing directly into each property. Many blocks were built up to three storeys high and each block may have an external communal area including sheds and drying area.

They were an ideal solution to the housing problem at the time and offered many variations, including a mix of maisonettes and flats. They are still a popular choice today.

Cantilevered Balconies

These are balconies constructed so that the balcony protrudes from the face of the building without any visible support. The projecting balcony was usually a direct extension of the floor slab. The loads are supported by cantilevering the structure off the wall. Later, many of these concrete balconies suffered from drainage and thermal bridging issues.

Right: Pitched roof low-rise block with concrete balconies.

'Coming Home to a Rationalised Traditional Home.'

During the 1960s and 1970s, some construction methods became less complicated in design and build. Financial savings were made on materials, labour and build time costs.

These methods were called Rationalised Traditional construction. Typically, only the side walls were masonry, forming a cross-wall construction. The front and rear elevations were infilled with timber framed panels and a variety of external wall designs. All of these non-masonry dimensions were standardised and often part assembled in factory units and then delivered to the site.

Right: Properties have been modernised with cavity wall insulation and insulation slabs installed in between the timber studs.

'Coming Home to a Flat Roofed Home.'

Non-traditional housing methods of construction favoured timber framed systems. These involved using timber wall panels to the inner skin, timber floor panels and an external skin using materials such as vertical tiles, timber or brick cladding.

Above: Bungalow with modernised pitched roof.

Flat roofs were built for cost efficiency but sadly failed. The poor insulation, ventilation or vapour barriers lead to heavy condensation. Large quantities had to be pitched roofed a decade or so later, as the condensation and leaks impacted on the timber structure.

Above right: Terraced bunaglows and garages.

'Coming Home to a House Alteration.'

The slum clearances that began during the Industrial Revolution had now managed to move 900,000 people out of slums during the 1950s and 1960s.

Due to the ongoing inflated building costs, scarcity of materials and skilled labour, many of the late Victorian terraced homes were modernised or bought for housing needs. They are a solid wall construction and often had cellars.

Alterations allowed for these properties to be modernised to standards of the present day and were often adapted or extended over three floors to cater for large families. Alterations may include extensions; loft conversions; moving a staircase; installing a roof light; installing inside toilet facilities and removing a wall or chimney.

Right to Buy

The introduction of the 'Right to Buy' under the Housing Act 1980 changed the balance of social housing forever. Following this policy, the strong period of social housing growth began to decline.

During this period, a large proportion of the quality housing stock was sold to those who qualified for the right to buy. This included a large number of family houses as opposed to flats. Consequently, this altered the balance of the type of social housing available.

Right: Victorian terraced homes.

'Coming Home to a House Conversion.'

Conversions are defined as a change in use or function in a building. For example, a commercial or industrial property converted to residential use.

Whether an old pub, a post office or a local shop, conversions allow for more homes. Modern day Building and Planning Control oversees the adaption and conversion processes. Conversions can offer alternatives for run-down shopping centres or unused buildings.

Building Homes for the Future

In modern times, social housing has suffered from little Planning system support, reduced public investment, and rising land and development costs.

The present housing crisis suggests that more government funding will be required to assist families to cover the cost of unaffordable private rents through benefit payments, rather than investing in new homes with guaranteed low rents.

Right:Large pub converted into flats.

<h1 style="text-align:center">'Coming Home to a Modern
Extra Care Home.'</h1>

Historically, housing provision for the elderly population had been funded by charitable means. By the late 1940s, it was believed that the local councils had the responsibility to provide housing and that purpose-built housing for the elderly was to be included in the social housing programme.

Modern extra care buildings offer a design that accommodates the changing needs of residents and is capable of transforming services as the residents age. Residents live in a self-contained flat but staff are usually available 24 hours per day to provide personal care and support services.

Modern extra care homes or retirement villages are available to older people in a variety of tenures including affordable rent, shared ownership and leasehold sale. They are designed, built and equipped to the highest specifications.

Right: Modern, timber-frame building with large communal areas.

'Coming Home to a Modern Methods of Construction Home.'

Modern Methods of Construction (MMC) allow for building components to be assembled or part assembled in factory conditions and offer an alternative to traditional construction. Modern Methods of Construction are bringing a revolutionary change to the building sector and address the skill and housing shortage in the UK. They offer speed and efficiency in house building.

There are many examples of MMC including timber, steel or concrete frames, modular volumetric systems and even 3D printing methods.

Volumetric modular units can be made as complete, single-storey units in factory conditions that can be linked together to form complete buildings.

Above: Assembly of ground floor and first floor pods.

The pods are pre-installed with electrics, plumbing, heating, doors, windows and internal finishes and can be delivered separately as ground floor, first floor, and second floor pods.

Above: Multiple assembled housing pod blocks.

'Coming Home to a Modern Timber Frame Home.'

Non-traditional methods of construction were replaced by traditional methods by the 1980s and remained so until the turn of the century.

Unlike a traditional brick and block home, a modern timber frame build can have large components prefabricated in factory conditions. Modern timber frame structures are considered a modern method of construction (MMC) and consist of a timber, load-bearing frame with an outer leaf of brick wall.

This modern method of constructing buildings incorporates advanced breathable membranes, insulation and vapour control layers along with careful detailing to ensure durability. Timber frame is one of the fastest methods of construction and the average home can be completed from concrete foundation slab to weathertight stage in about 5 days.

Some traditional cavity brick and block new-builds can incorporate large prefabricated components (eg. party walls) to form hybrid new-builds and reduce build times too.

Right: Scaffolding supporting the timber frame building as it is being constructed.